Who Am I?

Ocean Animals

Written by Read With You
Center for Language Research and Development

The characters and events portrayed in this book are fictitious. Any similarity to real persons, living or dead, is coincidental and not intended by the author.

Published by Read With You Publishing

Designed by Read With You Center for Language Research and Development

ISBN-13: 978-1-944710-65-1
ISBN-10: 1-944710-65-5

Printed in the United States of America.

I have four teeth and can puff up.

Who am I?

pufferfish

I sleep with one eye open.

Who am I?

dolphin

I am the largest animal on earth.

Who am I?

whale

I am a fish that looks like a snake.

Who am I?

eel

I have forty to forty-five teeth.

Who am I?

shark

I don't have a brain.

Who am I?

jellyfish

My flippers help me swim well.

Who am I?

sea turtle

I lose all of my teeth and scales by adulthood.

Who am I?

swordfish

Puffer Fish

Dolphin

Whale

Eel

I am a puffer fish.

I am a dolphin.

I am a whale.

I am an eel.

Shark
Jellyfish
Sea Turtle
Swordfish

I am a shark.

I am a jellyfish.

I am a sea turtle.

I am a swordfish.

We are ocean animals.

An ocean is a huge body of salt water. There are five large oceans in the world.